41st Biennial Meeting of the American Society of Sugar Beet Technologists 2021

Online
1 - 4 March 2021

ISBN: 978-1-7138-4276-7

Printed from e-media with permission by:

Curran Associates, Inc.
57 Morehouse Lane
Red Hook, NY 12571

Some format issues inherent in the e-media version may also appear in this print version.

Printed by Curran Associates, Inc. (2022)

American Society of Sugar Beet Technologists
800 Grant Street, Suite 300
Denver, Colorado 80203-2944

Phone: 303-832-4460
Fax: 303-832-4468

www.bsdf-assbt.org

Additional copies of this publication are available from:

Curran Associates, Inc.
57 Morehouse Lane
Red Hook, NY 12571 USA
Phone: 845-758-0400
Fax: 845-758-2634
Email: curran@proceedings.com
Web: www.proceedings.com

TABLE OF CONTENTS

AUTOMATED FACTORY EXCEPTION REPORTING USING SUGARS^{TM} 4.1 MASS AND ENERGY BALANCE SOFTWARE

Scott M. Kahre
Sugars International LLC, 305 Inverness Way South Unit 204, Denver, CO 80112

Introduction:

Sugar factory performance is often compared to a set of "standards" that are set at the beginning of each campaign based on anticipated average beet quality parameters. However, as beet quality and processing rates change, even the most optimized factory's results will vary. It is more beneficial to compare the actual factory operating data to an accurate mass and energy balance that accounts for the actual throughput and beet supply conditions. This will give a much more realistic picture of the factory's performance versus what would be expected under the actual conditions experienced.

Methods:

The first step to automated exception reporting is to build and validate an accurate mass and energy balance model of the factory. This involves a thorough audit of the entire process, including installed equipment, piping, manual valve settings, and typical process operating parameters. The model should be designed in such a way that process chemical inputs, steam loads, sugar and by-product production and other results are calculated based on inputs of beet slice rate, cossette sugar content, non-sugar load, beet temperatures, and other external variables.

Sugars^{TM} is the mass and energy balance software used throughout the global sugar industry. The figures below show an example beet sugar factory simulation model in Sugars^{TM}.

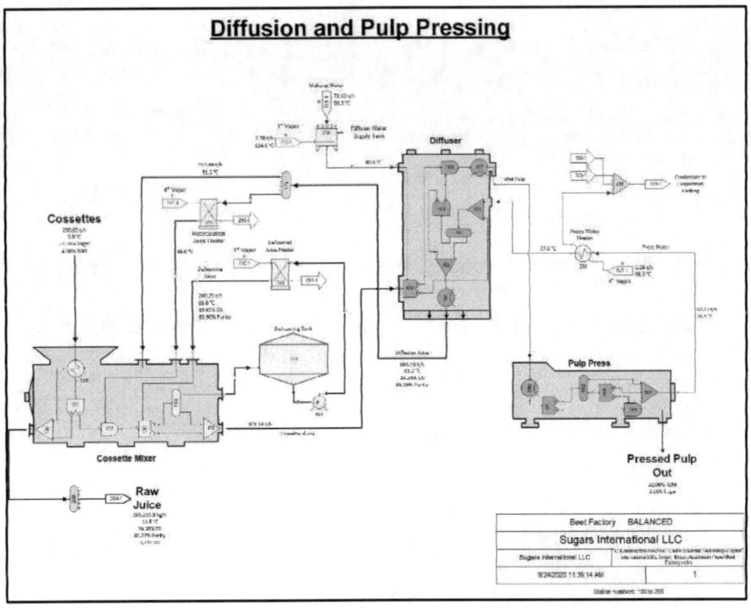

Figure 1: Beet Factory model, Diffusion Area

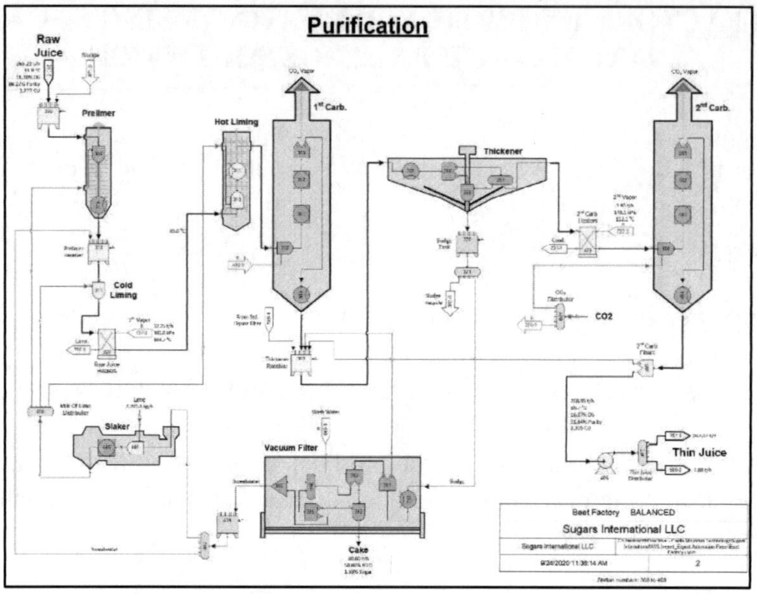

Figure 2: Beet Factory model, Purification Area

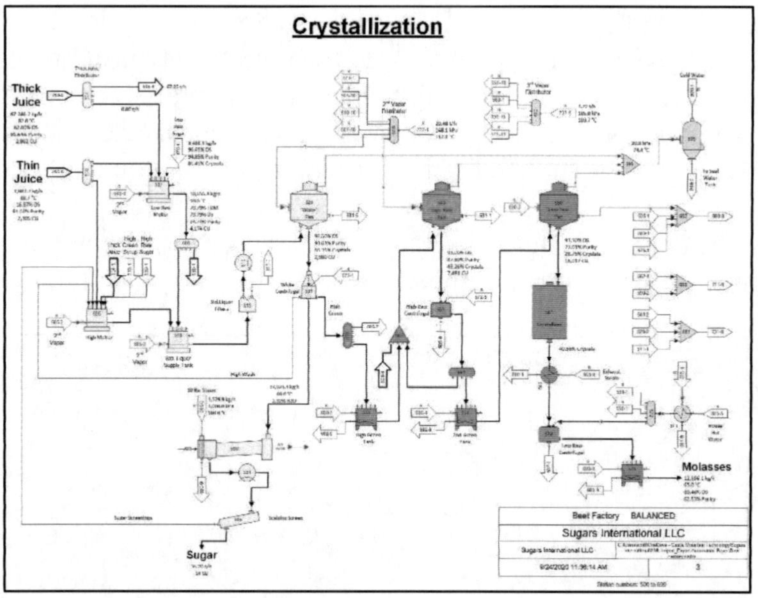

Figure 3: Beet Factory model, Crystallization Area

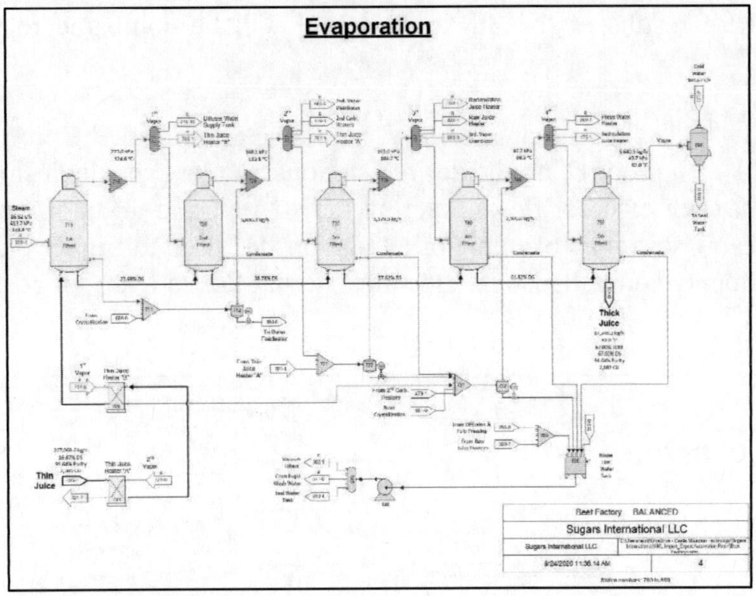

Figure 4: Beet Factory model, Evaporation Area

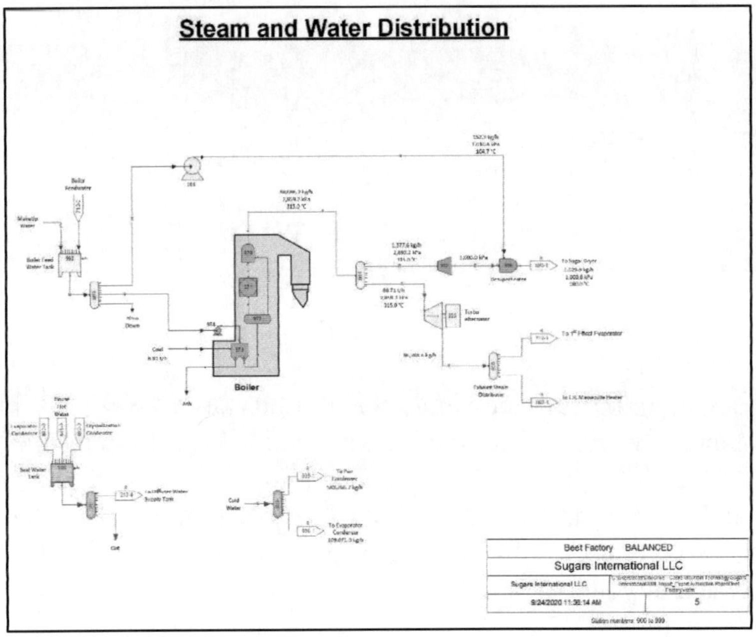

Figure 5: Beet Factory model, Steam and Water Area

Following model development, it is necessary to identify which process parameters will need to be included in the XML data import and export to and from Sugars™. For input to the Sugars™ model, the XML file should include, at a minimum, the external flow stream containing the cossettes entering the factory. This will allow parameters such as the beet slice rate, cossette sugar content, dissolved non-sucrose content, fiber (marc) content, and even cossette temperature to be input to the model. Depending on factory configuration, other input variables may include stored juice processing rates/quality, steam turbine generator loads, or other factory settings.

Additionally, flow streams containing sugar and co-product production rates and quality should be included in the XML file, as well as required energy and material inputs such as steam and limestone. These results of the Sugars™ model will be compared to the actual factory operation.

Including Sugars™ flow stream and station parameters in the XML import or export files is easy. Simply add a "Station ID" to the desired stations; that triggers inclusion in the XML file of that station's data, any external flows entering the station, and all internal flows leaving the station. Figure 6 below shows an example of a "Station ID" used to trigger inclusion of the first Cossette Mixer station, which will also include the external flow stream for cossettes entering the factory.

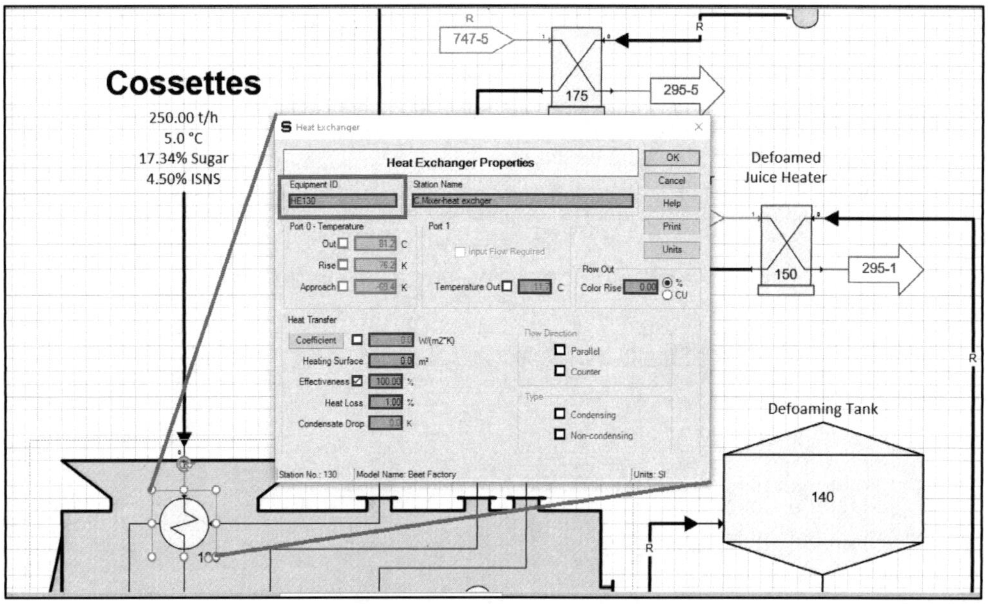

Figure 6: Adding "Equipment ID" to Sugars™ station

Once an "Equipment ID" is added to all the desired stations, an XML Data Export should be initiated to create an XML data file for the Sugars™ model. XML data files are a standardized text file format used in many program and web applications. Figures 7 and 8 below show the XML data export feature in Sugars™ and a portion of the XML data file for the Beet Factory example model given above. Note that to utilize this feature in Sugars™, the XML Data import/export feature must be activated in your software license.

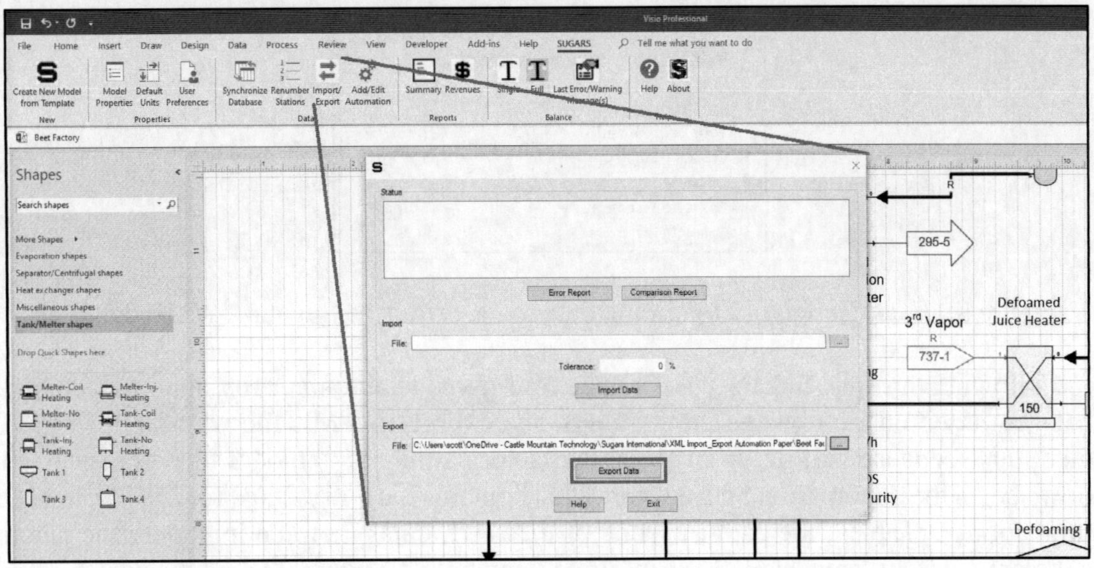

Figure 7: Manually exporting XML data file from Sugars^{TM}

```
<?xml version="1.0" encoding="UTF-8"?>
- <AMSDocument xmlns="http://sugarsonline.com/AMSSchema.xsd" Units="SI" Timestamp="2020-09-24T00:00:00"
  xmlns:xsd="http://www.w3.org/2001/XMLSchema" xmlns:xsi="http://www.w3.org/2001/XMLSchema instance">
  - <ExternalFlow>
      <EquipmentID>SlkrBlndr</EquipmentID>
      <PortNumber>1</PortNumber>
    - <Property Name="Pressure">
        <Value>101.400002646921</Value>
      </Property>
    - <Property Name="Temperature">
        <Value>15</Value>
      </Property>
    - <Property Name="FlowRate">
        <Value>7027.3524443107954</Value>
      </Property>
    - <Property Name="Water">
        <Value>0</Value>
      </Property>
    - <Property Name="EthanolLiquid">
        <Value>0</Value>
      </Property>
    - <Property Name="DissolvedSucrose">
        <Value>0</Value>
      </Property>
    - <Property Name="Invert">
        <Value>0</Value>
      </Property>
    - <Property Name="Ash">
        <Value>0</Value>
      </Property>
    - <Property Name="DissolvedNonSucrose1">
        <Value>0</Value>
      </Property>
      <Property Name="DissolvedNonSucrose2">
```


Figure 8: Example Sugars^{TM} XML data file structure (only first few lines shown)

This exported XML data file will serve as the template for development of the interface from/to the factory's laboratory and/or process information system.

Programming this interface is the next required step, and the method will vary depending on the factory's data system(s). In all cases, a program should be written that opens the XML data file, over-writes the pertinent input parameters (slice rate, cossette quality, etc.) with their actual average values, and saves the file in a location accessible to Sugars^{TM}. This program should execute automatically on a specific frequency of the factory's choosing (daily, weekly, semi-weekly, etc.). Keep in mind that averages over at least a week's time will likely be more representative than more frequent aggregates due to fluctuations in factory tank levels and other transient conditions. A relatively simple example of factory operating data to be written from Microsoft Excel to the Sugars^{TM} XML file is shown in Figure 9 below.

	A	B	C	D	E
1	Beet Factory Daily Operating Report				
2					
3	Input Values				
4	Tons Beets Sliced:			5,482	tons
5	Cossette Sugar Content:			16.85	%
6	Cossette Purity:			86.2	%
7	Cossette Temperature:			2	deg. C
8	Cossette Marc:			4.5	%

Figure 9: Example factory operating data to be written to Sugars™ XML data file

Following this step, Sugars™ 4.1 can be configured to automatically import this XML data file, balance based on the updated input values, and export an XML file with the results. Once activated, this sequence will occur any time the factory's Sugars™ model file is opened. Figure 10 below shows the Automation Settings form that contains the entry fields for specifying whether the automation is enabled, the path/filename of the XML data file to be imported, the parameter change tolerance (to trigger internal exception logging within Sugars™), and the path/filename of the XML data file that Sugars™ will export with the balance results. Note that once Automation Settings are configured, a text document with filename [modelname].txt will be saved in the same directory as the Sugars™ drawing and model database files. This file must remain in that directory for the automation process to proceed.

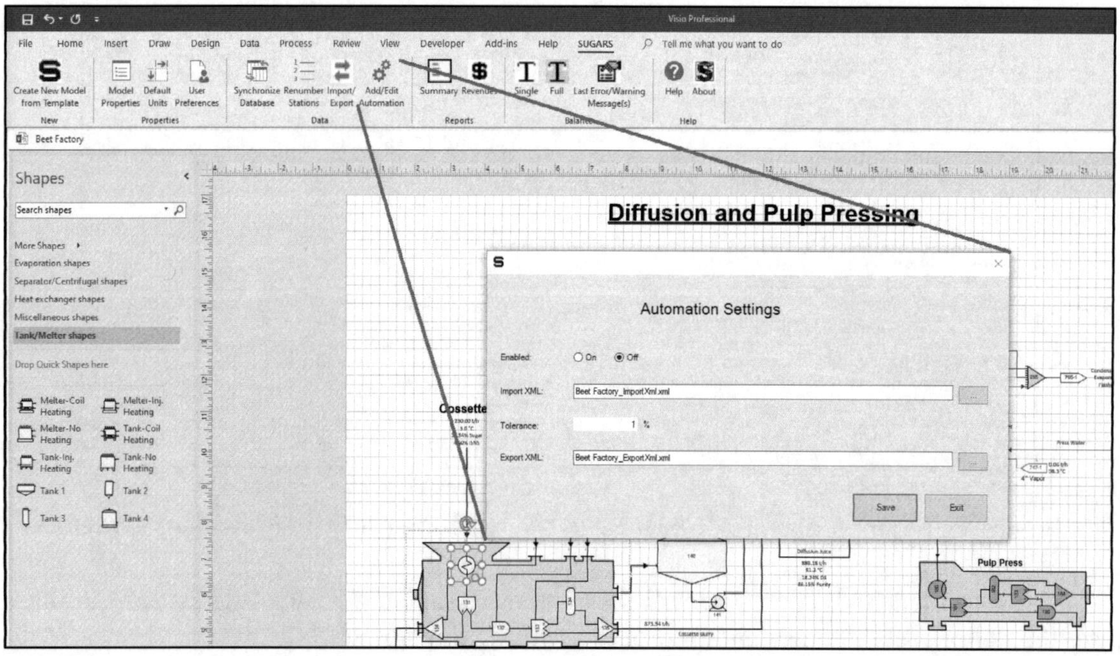

Figure 10: Sugars™ XML Import/Export Automation Settings

The factory's Sugars™ model can be scheduled to automatically open on a specified frequency using Windows Task Scheduler. This will cause the model drawing to open, launch Sugars™, import the XML data file containing input data, balance the model, then export the XML data file with the results, and finally close Visio/Sugars™. The final screen of the "Create Basic Task Wizard" in Windows Task Scheduler is shown below in Figure 11, showing an example setup for automatically opening the Sugars™ model. Note that the "Argument" field is not shown, but must be set to the full path/filename of the Visio drawing file to be opened.

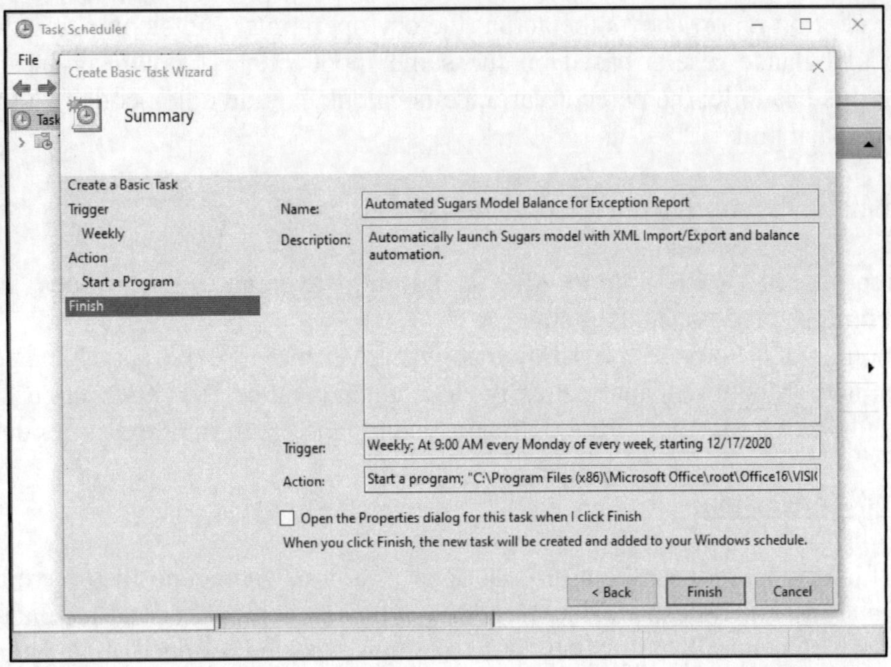

Figure 11: Example Windows Task Scheduler settings

With the Sugars™ model fully automated, the final step is to build an automated report that reads the Sugars™ balance results from the exported XML data file, and compares them to the actual factory operating results. This could be built as a Microsoft Excel spreadsheet using Visual Basic for Applications macros, or using SQL database programming, or any other platform based on local preference. A simple example is shown in Figure 12 below using Excel.

Beet Factory Daily Operating Report Comparison					
Input Values					
Tons Beets Sliced:	5,482	tons			
Cossette Sugar Content:	16.85	%			
Cossette Purity:	86.2	%			
Cossette Temperature:	2	deg. C			
Cossette Marc:	4.5	%			
Output Values	**Report**		**Sugars**		**Variance**
Lime Usage:	175	tons	169	tons	3.8%
Thick Juice Purity:	91.8	%	91.6	%	0.2%
Sugar Production:	745	tons	770	tons	-3.2%
Molasses Production:	271	tons	275	tons	-1.3%
Molasses Purity:	63.4	%	62.5	%	1.4%
Steam Usage:	1,822	tons	1,939	tons	-6.0%
Steam/Beets:	33.2	%	35.4	%	-6.0%
Pressed Pulp Production:	661	tons	672	tons	-1.7%
Pressed Pulp Moisture:	77.5	%	77.2	%	0.3%
Pulp Loss:	0.28	%/bts	0.26	%/bts	6.6%
Total Extraction:	80.7	%	83.3	%	-3.2%

Figure 12: Example Exception Report

This exception report shows the actual factory operating results compared directly to the Sugars^{TM} model balance results based on the same input values (shown again at the top for reference). In this example, the percent variance is calculated and color-coded so that any value over +/-5% is shaded pink.

In summary, the entire sequence of automation is as follows:

1) Creation of XML import file containing average input values from factory laboratory/process data system.
2) Opening of Sugars^{TM} model drawing, with XML import/export and balance automation enabled. Model will automatically close when balance and export are complete.
3) Reading of XML export file (balance results) and creation of exception report.

Discussion and Conclusion:

Sugar factory operation exception reports are not a new concept; they have been utilized for many years across the industry. However, they typically compare actual results to a static set of "standards" which generally only apply to the average expected results over an entire campaign. Thus, on a more immediate basis (i.e. weekly), they are considered less applicable. For instance, an exception report that compares the operation of a factory in February conditions (potential for some frozen beets, some with moderate deterioration, reduced slice rate, etc.) to campaign average conditions is often ignored, as nearly every parameter shows major deviation.

However, when the basis for comparison is a Sugars^{TM} mass and energy balance model that accounts for these changes in factory inputs, the exception report becomes much more valuable. Factory staff can use the results to identify potential opportunities with regard to operational efficiency, measurement accuracy, and other issues needing attention. In many cases, addressing these items results in many thousands of dollars of increased revenue or reduced cost.

The additional XML Import/Export and balance automation feature in Sugars^{TM} 4.1, when combined with factory laboratory/process data systems, opens the door to automatic generation of useful exception reports on a regular basis. This is a valuable new tool for sugar companies across the globe to maximize revenues and minimize costs.

BLOOMQUIST MARK W.,* DAVID C. METTLER, and CODY J. GROEN. Southern Minnesota Beet Sugar Cooperative, 83550 County Road 21, Renville, MN 56284. **Methods to reduce *Cercospora beticola* inoculum in field.**

Cercospora leaf spot can be a devastating disease for sugar beet production in southern Minnesota. Since 2016, cercospora leaf spot levels have been high across the Southern Minnesota Beet Sugar Cooperative growing area. Six to seven fungicide applications are being applied to manage disease levels in fields which is placing a significant amount of selection pressure on these fungicides. As fungicide resistance trends upward, additional tools are necessary to manage this disease. In 2019 and 2020, SMBSC conducted a trial to look at potential methods to decrease the levels of *Cercospora beticola* inoculum that survive from one season to another in the soil and plant debris. Trials were established on areas that were planted to sugar beet the previous year and had high levels of cercospora present. There were five treatments in the trial each season. The treatments were tillage, heat/burning, copper fungicide application, hydrogen peroxide/peroxyacetic acid application, and an untreated check. The treatments were applied in the spring prior to planting. The chemical treatments were applied to the soil and crop residue present on the soil. Sugar beets were no-till planted across the trial area following the treatment application. The sugar beets were allowed to develop through the growing season, and when CLS began developing across the trial, foliar ratings were taken using the KWS 1-9 scale. The foliar ratings were conducted 2-3 times per week for a three week period. In the 2019 season, both the heat/burn treatment and the tillage treatment had lower average CLS ratings than the untreated check. These differences were statistically significant at the 0.05 significance level in comparison to the untreated check. In 2020, the heat/burn treatment was statistically lower in average CLS ratings versus the untreated check. The tillage treatment was numerically lower, but not significantly lower at the 0.05 significance level in 2020. The two spring applied chemical treatments did not lower the level of CLS present in either year of the trial. This trial will be repeated for a third year during the 2021 growing season.

LAMB, JOHN A.[1*], RYAN KOWTA[2], ISRAEL SANTIAGO[2], and MARK W. BLOOMQUIST[3]. [1]Univ. of Minnesota, 2182 Woodbridge Way, Woodbury, MN 55125, [2]Spreckels Sugar Company, 95 West Keystone Rd. Brawley, CA 92227, and [3]Southern Minnesota Beet Sugar Cooperative, 83550 County Road 21, Renville, MN 56284. **At What Plant Stand is Replanting Needed in the Imperial Valley of California?**

Justification: Establishing an optimum sugar beet plant stand is important to maximizing extractable sucrose yield in the Imperial Valley of California. If conditions are right at planting, then an optimum stand is the result. If there are problems such as seed quality, hot weather, or crusting soil, the stand will be reduced. Currently, the growers in the Imperial Valley aim for a plant stand of 30 sugar beet plants per ten feet of row for optimum production. At what plant stand should a grower decide to replant when the emergence is sub-optimal? The current answer to the question is 15 to 16 sugar beet plants per 10 feet of row. Since the advent of glyphosate resistant sugar beet varieties, there has been no research information from the Imperial Valley about replanting thresholds.

Objective: Determine the threshold plant stand that requires replanting in early and late harvested sugar beet production.

Methods and Materials: The treatments are listed in Table 1. The study was a randomized complete block design with 4 replications. There were seven plant stand treatments to represent different plant population thresholds. Stands were thinned after emergence. Treatments 1, 2, 3, and 4 were thinned to an even stand while treatments 5, 6, and 7 were thinned unequally simulating an uneven stand that would occur with poor emergence (gappy). Treatment 8 was planted later at a plant stand of 30 plants per 10 feet of row, similar to when a replant decision was made. Seven sites were established during the study from Fall 2017 to Summer 2020. Two of those sites were abandoned because of poor stands, Table 2. There were two early harvest and three late harvest sites. Sugar beet for early harvest in the Imperial Valley are usually planted in September and harvested between April 1 and early June. Late harvest sugar beet fields are usually planted in October and harvested in late June through the end of July. The planting, thinning, and harvest dates are listed in Table 2. All locations were planted with Beta 5460. Root yield was determined on the dates reported in Table 2. Root quality was determined by the Spreckels Sugar Tare Laboratory.

Table 1. Treatments for the proposed reduced stand study.

Treatment	Sugar beet per 10 feet of row
1.	34
2.	30
3.	26
4.	22
5.	18 gaps
6.	14 gaps
7.	10 gaps
8.	30 replant – planted at 4-5 weeks after original planting.

Table 2. Planting, thinning, and harvest dates for the replanting study.

Site	Planting date	Thinning date	Replant	Harvest date
Site 1 2017-2018	Oct. 2, 2017	Oct. 30, 2017	Nov. 15, 2017*	June 5, 2018 (early)
Site 2 2017-2018	Oct. 6, 2017	Nov. 9, 2017	Nov. 13, 2017	July 17 and 18, 2018 (late)
Site 1 2018-2019	Oct. 19, 2018	Nov. 27, 2018	Dec. 12, 2018	July 15, 2019 (late)
Site 1 2019-2020	Sept 14, 2019	Oct 17, 2019	Nov. 1, 2019	April 06, 2020 (early)
Site 3 2019-2020	Oct. 12, 2019	Nov. 11-12, 2019	Dec. 6, 2019	June 17, 2020 (late)

* birds ate seed from the previous planting in early November.

Results: <u>Root yield</u> Sugar beet root yield ranged from 36.8 to 92.4 ton per acre at the five sites, Table 3. The root yield was significantly affected by the plant stand treatments at three of the five sites, Table 3. The optimum yield at two sites, Late 17-18 and Late 19-20 were at 14 and 22 plants per 10 feet of row. The response at the Late 18-19 site was linear and the optimum was not obtained. The greatest root yield at the Late 18-19 site occurred at 34 plants per 10 feet of row.

Table 3. Sugar beet root yield as affected by plant stand from 2017 to 2020 in the Imperial Valley of California.

Stand	Early 17-18	Late 17-18	Late 18-19	Early 19-20	Late 19-20
Plants/10 ft. of row	Root yield (ton/A)				
10	61.2	94.4	70.2	36.1	48.2
14	63.9	96.6	77.3	35.8	53.9
18	63.3	92.3	72.9	38.2	56.2
22	64.5	95.1	77.2	35.2	58.5
26	65.2	92.4	74.4	38.6	56.4
30	63.6	90.7	77.7	37.2	55.1
34	61.4	85.9	78.9	36.7	57.3
Grand mean	63.3	92.4	75.5	36.8	55.1
Trt	NS*	0.02	0.03	NS	0.01
C.V. (%)	5.4	3.6	4.6	7.7	5.9

* NS = Not significant at the P > 0.05.

<u>Extractable sucrose per ton</u> Extractable sucrose per ton was only affected by the plant stand at one site, Late 18-19, Table 4. The significance for the Late 18-19 was caused by the increased extractable sucrose per ton at 26 plants per 10 feet of row. In most cases, quality as measured by extractable sucrose per ton is not affected by plant stand.

Table 4. Extractable sucrose per ton as affected by plant stand from 2017 to 2020 in the Imperial Valley of California.

Stand	Early 17-18	Late 17-18	Late 18-19	Early 19-20	Late 19-20
Plants/10 ft. of row	Extractable sucrose (lb/ton)				
10	279	240	279	274	311
14	280	240	286	277	316
18	278	259	285	276	310
22	286	244	284	280	316
26	292	249	297	272	322
30	273	238	285	274	322
34	281	244	286	279	314
Grand mean	281	245	286	276	316
Trt	NS*	NS	0.002	NS	NS
C.V. (%)	2.9	4.2	1.5	2.7	2.1

* NS = Not significant at the P > 0.05.

<u>Extractable sucrose per acre</u> Extractable sucrose per acre was significantly affected by plant stand at four of the five sites, Table 5. The effect of plant stand on extractable sucrose per acre from all sites is shown in Figure 1. In all cases except for the Early 19-20 site, the optimum stand ranged from 18 to 34 plants per 10 ft of row.

Table 5. Extractable sucrose per acre as affected by plant stand from 2017 to 2020 in the Imperial Valley of California.

Stand	Early 17-18	Late 17-18	Late 18-19	Early 19-20	Late 19-20
Plants/10 ft. of row	Extractable sucrose (lb/acre)				
10	17044	22682	19596	9878	15077
14	17893	23162	22086	9882	17034
18	17568	23918	20742	10544	17379
22	18388	23247	21932	9818	18452
26	19031	22975	22050	10597	18154
30	17335	21531	22102	10201	17752
34	17238	21169	22644	10225	17977
Grand mean	17785	22640	21578	10151	17394
Trt	0.04	0.02	0.02	NS*	0.001
C.V. (%)	4.8	4.3	5.0	6.6	5.4
Trt *	9631	17595	7679	5000	9850

* NS = Not significant at the P > 0.05.

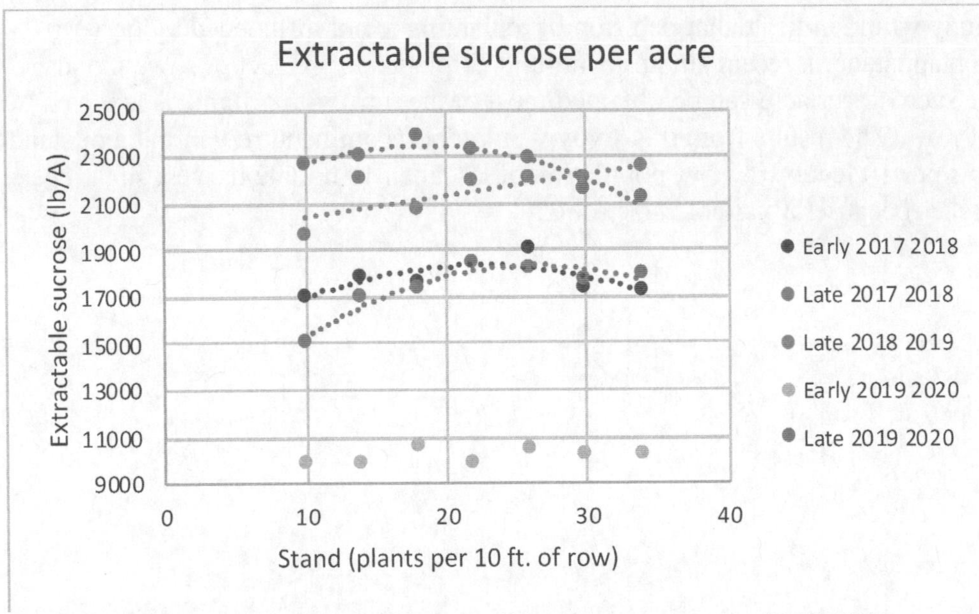

Figure 1. Plant stand effects on extractable sucrose per acre 2017-2020.

The economic effect of whether or not to replant will depend on income loss from later planting date and cost of added inputs. The effect of time between original planting date and the replanting date on extractable sucrose per acre was measured by comparing treatment 2, plant stand of 30 plants per 10 feet of row with the late replant treatment 8, Table 6. The average extractable sucrose per acre loss in this study was 160 lb per acre per day or 1.00% per day.

Table 6. Optimum plant stand for extractable sucrose per acre at five locations in the Imperial Valley from 2017 to 2020.

	Optimum stand	Days after planting	Trt 2 – Trt 8	Replant extractable sucrose per acre losses based on Trt 8.	
Site	Plants/10 feet of row	Days	lb/A	lb extractable sucrose per acre per day	Loss per day of planting delay (%)
Early 17-18	22	44	7704	175	-1.01
Late – 17-18	18	38	3936	104	-0.48
Late – 18-19	34	54	14424	267	-1.21
Early – 19-20	No optimum	48	5201	108	-1.06
Late – 19-20	26	55	7902	144	-0.81

Besides the loss in extractable sucrose per acre, the addition cost of replanting (fuel, seed, time, etc.) needs to be considered. One other factor is the harvestability of sugar beet. The size of the some of the sugar beet roots at harvest in the uneven 10 plants per 10 feet of row treatment (1) were very large and caused significant mechanical harvest issues. This was particularly an issue with the late harvest sites.

Summary: This study would indicate that decision of replanting is not influenced by harvest date. The optimum plant stand in recent studies is around 24 plants per 10 feet of row. Good yields of extractable sucrose per acre can be obtained most of the time with a stand as low as 14 plants per 10 feet of row. The results from this study would not recommend replanting a of stand greater than 14 plants per 10 feet of row. A population of less than 14 on late harvest sugar beet can cause harvest issues because of variable size roots.

Investigations of Cercospora leaf spot Biology in Sugar Beet
By: Amanda R. Koch and W.L. Stump

The relative fitness of benzimidazole and strobilurin insensitive *Cercospora beticola* isolates were compared to a fungicide sensitive isolate on sugar beet in the greenhouse at the University of Wyoming. Disease development was determined on both a Cercospora leaf spot resistant and susceptible variety. The sensitive and insensitive isolates were recovered from sugar beet fields from the Western Sugar production region. After inoculation, parameters of lesion development, expansion and sporulation were measured up to 24 days post inoculation. Lesion numbers and expansion for those isolates were in general greater in the susceptible variety compared to the resistant. Lesion expansion showed differences in the isolates in how they responded to the cultivars. Disease severity (0-100) showed slight differences between isolates. Sporulation measurements were variable for the susceptible cultivar and consistent at zero for the resistant cultivar across all three isolates. This study is ongoing and currently in its second round at the university's greenhouse.

Development of U.S. BNYVV infectious clones to study Rz1 and Rz2 resistance breaking.

Alyssa Flobinus, John J. Weiland and Melvin D. Bolton

Beet necrotic yellow vein virus (BNYVV) is a *Benyvirus* consisting of four positive-sense single-stranded RNAs. BNYVV is the main agent responsible for the most devastating sugar beet viral disease worldwide, Rhizomania. Rhizomania leads to a major reduction in sugarbeet yield and sugar content, resulting in significant financial losses to the industry. To manage BNYVV, three sugar beet genotypes possessing Rz1 (the mostly widely used resistance gene), Rz2, or more recently the combined Rz1+Rz2, have been deployed. Over the last 16 years, resistance-breaking (RB) strains of the virus have been reported for Rz1, a phenomenon associated with modifications in the gene p25 encoded on RNA3. More recently, RB strains have been identified that appear to overcome Rz2- and Rz1+Rz2-mediated resistance, but the mechanism of RB remains elusive. In order to investigate the role of RNA-3 and potentially other genome regions in the RB phenomena, BNYVV infectious clones have been created for the first-time using U.S. strains of the virus.

BNYVV RNA-1 and -2 clone design and construction was based on RNA sequencing data obtained from BNYVV-infected plants grown in infested soil possessing RB isolates of virus. RNA-3 and -4 clones were directly isolated from these plants *via* RT-PCR. Our results shown that these clones, when transcribed to produce capped RNAs, are infectious on *Chenopodium quinoa* and *Beta macrocarpa* hosts. Protein p25 modifications associated with RB in U.S. will be introduced in the p25 gene and tested in sugarbeet genotypes encoding the different resistance gene complements. Through use of these clones, a better understanding of BNYVV virulence in general and the RB phenomenon specifically will be obtained. Additionally, the clones will allow us to study other viral functions, some of them may determine virus transmission and aggressiveness, leading to strategies for the inhibition of BNYVV infection in plants.

Keywords: Resistance, BNYVV, Infection, Gene.

STILES II, BRIAN. Michigan State University, 7955 S Durand Rd. Durand, MI 48429 **Cereal rye suppresses glyphosate-resistant horseweed in sugarbeet**

Glyphosate-resistant horseweed proves to be a challenge for many Michigan sugarbeet growers. Integrating multiple management strategies, including a cereal cover crop, may improve horseweed control. In 2019 and 2020 field studies were conducted in East Lansing, Michigan to evaluate the effects of fall-planted cereal rye termination time and method in combination with different postemergence (POST) herbicide treatments for horseweed control. Cereal rye was drilled at 67 kg ha-1 in the fall of 2019 and 2020. The studies were conducted as a split-plot design with cereal rye termination method and time as the main plot factors and herbicide treatments as the sub-plot factor. Cereal rye treatments included: early burndown (EBD) 14 d prior to sugarbeet planting, burndown at planting (PBD), PBD + roller, and PBD + roller crimper, a delayed burndown (DBD) 14 d after planting, and a no cover control, the addition of no-cover strip-till and delayed burndown strip-till were treatments added to the 2020 study. The burndown treatment consisted of glyphosate applied at 1.27 kg ae ha-1 + ammonium sulfate. The three herbicide treatments consisted of two POST applications at the 2- and 6-8 leaf sugarbeet stage. The treatments included: 1) glyphosate twice (control), 2) glyphosate (0.84 kg ae ha-1) followed by glyphosate (0.84 kg ae ha-1) + clopyralid (0.11 kg ha-1), and 3) glyphosate (0.84 kg ae ha-) + clopyralid (0.06 kg ha-1) followed by glyphosate (0.84 kg ae ha-1) + clopyralid (0.11 kg ha-1). In 2019, cereal rye biomass at the time of termination was 640 and 740 kg ha-1 for the EBD and PBD terminations, respectively. Cereal rye biomass at the time of the 'Planting Green' termination was 5-times higher (4,200 kg ha-1). Horseweed biomass 14 d after planting (DAP) was 11 times lower where a cover crop was planted compared with the no cover control, regardless of termination time or method. 'Planting Green' with an application of clopyralid applied either once or twice reduced horseweed biomass up to 99%. At harvest, the main effect of the cereal rye cover crop reduced horseweed biomass up to 75%. The main effect herbicide treatments showed a greater reduction in horseweed biomass with two applications of clopyralid followed by clopyralid one application, followed by the control. Even though horseweed biomass was lowest in the DBD treatment, sugarbeet yield was reduced and was not different compared with the no cover control, due to reduced sugarbeet growth in the DBD. Sugarbeet yield for the EBD, roller crimper, and roller were all similar and the PBD showed the highest overall sugarbeet yield and was not different than the roller treatment. Regardless of clopyralid treatment, sugarbeet yields were the same. Data for the 2020 season is currently being analyzed. However, it appears that integrating cereal rye into a sugarbeet system could be an important tool for horseweed management. This research will be repeated in 2021.

A new portable qPCR technology to identify *Cercospora beticola* in asymptomatic leaves of sugar beet

Authors:

Jon Neubauer[1], Mark Bloomquist[2], and Melvin Bolton[1]

[1]United States Department of Agriculture, Fargo, ND.

[2]Southern Minnesota Beet Sugar Cooperative, 83550 County Road 21, Renville, MN. 56284.

Abstract:

Cercospora leaf spot (CLS) of sugar beet is caused by the fungal pathogen *Cercospora beticola*. Management of CLS relies on timely fungicide applications. Efficacy of most fungicides is enhanced when applied prior to *C. beticola* infection. However, host genetics and environmental dynamics make it difficult to predict when *C. beticola* will infect sugar beet in the field. Consequently, there is a need for diagnostic tools that are highly sensitive, specific, and ideally are portable so analyses can be carried out in the field. In this study, we used a portable magnetic induction cycler (MIC) and qPCR probes specific to *C. beticola* to identify the pathogen in asymptomatic leaves harvested from growers fields in southern Minnesota. The MIC has the capability of detecting fluorescent probes during short run times (under 50 min). Starting in the first week of June, leaf samples were harvested from sugarbeet fields over a five-week period. Detection occurred in the first week (10%) and increased throughout the sampling (34%). Fields with early detection eventually (within 14-21 days) had visual symptoms. Consequently, knowledge of asymptomatic *C. beticola* infection informed growers that fungicides should be applied as soon as possible to help minimize the impact of CLS. Since MIC can detect a range of probes, future work utilizing this machine to detect fungicide-resistant *C. beticola* isolates is under development. Taken together, recent advances in portable qPCR technology has laid the groundwork for highly sensitive detection of *C. beticola.*

EPIDEMIOLOGICAL STUDIES OF CERCOSPORA LEAF SPOT OF SUGAR BEET FOR IMPROVED MANAGEMENT

Alexandra P. Hernandez*[1], Daniel M. Bublitz[3], Tom Wenzel[3], Chris Bloomingdale[1], Cameron Pincumbe[1], Cheryl L. Trueman[4], Linda E. Hanson[2], and Jaime F. Willbur[1]; [1]Michigan State University, Department of Plant, Soil and Microbial Sciences, [2]Michigan State University, Department of Plant, Soil and Microbial Sciences; United States Department of Agriculture – Agricultural Research Service, [3]Michigan State University, Department of Plant, Soil and Microbial Sciences; Sugarbeet Advancement, [4]University of Guelph – Ridgetown, [1]612 Wilson Rd. Rm 35 East Lansing, MI 48824, [2]612 Wilson Rd. Rm 37 East Lansing, MI 48824

Cercospora leaf spot (CLS), caused by the fungal pathogen *Cercospora beticola*, is a major foliar disease of sugar beet in the Great Lakes region. In Michigan and Ontario, Canada, CLS management has proved difficult with increased *C. beticola* inoculum survival following high disease pressure in the field. Recent studies have indicated a potential for underprediction in certain disease models, suggesting that recommended spray timing may need revision for current fungicide usage. In this research, methods to reduce inoculum survival and ways to incorporate spore presence into prediction models were investigated.

Spore abundance and environmental factors were monitored in two locations in 2019 for use in model development. Using this data, a preliminary model ($R^2 = 0.23$, $P < 0.0001$) was developed to predict daily *C. beticola* spore abundance using daily precipitation ($r = 0.22$, $P = 0.05$), minimum daily relative humidity ($r = 0.24$, $P < 0.01$), maximum daily soil temperature ($r = -0.24$, $P < 0.01$), and maximum daily wind speed ($r = 0.38$, $P < 0.0001$). *Cercospora beticola* spore presence and abundance were monitored using mechanical and live spore traps in multiple locations in Michigan and Ontario, Canada over two years. Environmental factors were tested for correlations to spore abundance and used to model spore counts. Environmental variables with notable correlations to spore abundance include maximum and minimum air temperature (r = 0.35, $P < 0.0001$; r = 0.33, $P < 0.0001$), minimum and maximum soil moisture (r = 0.22, $P = 0.0132$; r = 0.22, $P = 0.0136$), maximum wind speed (r = 0.17, $P = 0.0630$), and precipitation (r = 0.12, $P = 0.1099$). Additional locations and years will be added to these observations for further model refinement and validation. A preliminary model will be used in field validations conducted in 2021.

Along with developing a preliminary model to predict *C. beticola* spore abundance, reducing the amount of inoculum before the growing season could improve CLS management. In this study, three fall treatments aimed at inoculum reduction, including plowing, foliar heat treatment, and foliar desiccation, were compared to a non-treated control. After end-of-season treatments, leaf samples taken at harvest were evaluated and leaf samples left in the field were assessed 45-, 90, and 135-days to determine *C. beticola* viability. During the following season, susceptible beets were planted in the same plots, and disease ratings were recorded and lesions on sentinel beets were evaluated to determine inoculum levels present. In 2019, significant treatment differences were detected in percentages of lesion sporulation ($P < 0.0001$) and lesion viability ($P < 0.05$) in at harvest samples (N=133 leaves and 240 lesions) with notable reduction in the heat-treated samples. The heat-treated samples had a 71% reduction in mean sporulation and a 35% reduction in lesion viability compared to the nontreated control. In 2020, reduced numbers of CLS lesions were observed on sentinel beets (N = 1 box of 4 sugar beets per plot) collected in 2019 heat treated plots from May 26-June 1 ($P < 0.05$) and June 2-9 ($P < 0.01$). The heat treatment also significantly reduced the area under the disease progress curve (AUDPC), calculated from severity ratings collected from 16 June to 28 July the following season after treatment application ($P < 0.01$). Mean AUDPC was 8.5 for heat treated plots compared to 19.9 for the nontreated control. In the repeated trial initiated in 2020, lesion sporulation was reduced in at harvest ($P < 0.0001$), 45-days post-harvest ($P < 0.01$), and 90-days post-harvest samples ($P < 0.05$) with significant reductions in heat treated plots. The at-harvest samples for the 1 mph and 3 mph heat treatment showed a 78% and 70% reduction in sporulation, respectively, compared to the nontreated control. Both heat treatments and the plow treatment all had 0% mean sporulation compared to 22% and 7% mean sporulation in the nontreated control for the 45- and 90-days post-harvest samples, respectively. Percent leaf degradation was calculated using initial and final leaf weights for overwintering samples (N = 160 leaves for each timepoint). No significant differences were detected between treatments for percent leaf degradation of at harvest or 45-day post-harvest leaf samples ($P > 0.05$). Significant treatment differences were detected for percent leaf degradation in 90-days post-harvest samples ($P < 0.02$). The plow treatment (83% leaf degradation) and the 1 mph heat treatment (82% leaf degradation) had the highest numerical percentages of leaf degradation for 90-days post-harvest samples. Remaining leaf samples from inoculum overwintering studies will continue to be evaluated. In 2021, early-season spore presence and abundance, weekly disease ratings, and final yield and sugar data will be collected to validate the efficacy of inoculum reduction strategies. The overwintering study has shown that the heat treatment, applied with a foliar burner at harvest, significantly reduced *C. beticola* sporulation, lesion viability, and the amount of disease the following season as compared to the other treatments. Therefore, heat treatment using a tractor-mounted foliar burner has the potential to significantly reduce inoculum levels and aid in CLS management in the field. These novel management practices will be further validated and could aid in sustainable management of CLS.

HENDERSHOT CARLY N.[1] CHRIS BLOOMINGDALE[1], THOMAS R. GOODWILL[2], CAMERON PINCUMBE[1], RANDOLPH BEAUDRY[1], LINDA E. HANSON[2] and JAIME F. WILLBUR[1], [1]Plant Biology Laboratory, 612 Wilson Road, Room 35 East Lansing, MI 48824, [2]Plant Biology Laboratory, 612 Wilson Rd. Room 37 East Lansing, MI 48824.

Characterization of pathogens and foliar disease interactions impacting sugar beet storage

In the US, sugar beet (*Beta vulgaris*) production has a market value of over $5 billion. In Michigan, sugar beets are stored for up to 200 days post-harvest while awaiting processing. Beets are stored primarily in covered or uncovered outdoor piles, although there is one indoor piling facility in Sebewaing, MI. During this storage period there are several factors that cause the beets to rot, reducing the sugar content and overall profit. In addition to root degradation, fungal mycelium, bacterial lipopolysaccharides, and pectinase byproducts clog filters and slow the extraction process. Michigan Sugar estimates that delaying onset of storage rot one month could save the industry $1 million per year. Proper identification and characterization of the pathogens affecting stored beets will allow for better targeting of disease management. In addition to organisms that directly colonize the beet root, there has been conflicting reports on how foliar disease in the field will affect the storability of the beet.

Cercospora leaf spot (CLS) is the most damaging and widespread foliar disease of sugar beet. To investigate the impact CLS has on pathogens colonizing the beet root, post-harvest trials will evaluate symptom development after inoculation with *Fusarium graminearum, Botrytis cinerea, Penicillium vulpinum*, and *Phoma betae* in beets with high or low in-season CLS severity. Beet root rot will be rated monthly between harvest and 150 days into storage. Preliminary research in North Dakota suggests that CLS does not affect the physiological storage qualities of the beet (Fugate et al., 2021). In contrast, Smith and Ruppel (1971) reported CLS could lead to increased storage rot, although no specific information was given about pathogens involved. This study will also examine varietal effect on rate of rot using CLS resistant and susceptible varieties. In addition to root rot, respiration rate will be also be monitored in beets with high and low CLS levels.

During storage in 2019 and 2020, *Penicillium* spp. (55%), *Botrytis cinerea* (12%), and *Fusarium* spp. (11%). were most frequently isolated from Michigan piling grounds. Other fungal pathogens isolated from both indoor and outdoor storage piles included *Geotrichum* spp., *Alternaria alternata,* and *Aspergillis niger*. The *Fusarium* spp. will be characterized to the species level by a combination of genotyping and morphology. *Geotrichum* spp. were isolated from the field in fall of 2019 and was found in storage in 2020. *Geotrichum* spp. has not previously been reported on beet in Michigan. *Geotrichum candidum* was recently reported on sugar beet in North Dakota and Minnesota (Khan et al. 2020); this fungus causes rubbery or sour rot on potato and is often characterized by a sour odor. The field isolate from 2019 is being used in the storage trials this year. Future goals include determining the virulence, and epidemiology of these storage pathogens to make management recommendations help reduce infection.

Two storage trials were conducted during winter of 2020 that looked at factors impacting the susceptibility of sugarbeet to postharvest disease. Trial 1 examined the impact of CLS infection in the field, and trial 2 evaluated the impact of variety in addition to CLS infection. Trial 1 used C-G333NT, and trial 2 used sugarbeet varieties C-G333NT and F1042 (Campbell, 2015) as CLS-susceptible materials and HIL-9865 and EL50/2 (McGrath, 2012) as CLS-resistant materials. Results from the storage trial 1 showed no significant differences between storage rot susceptibility in beets with high or low CLS levels in the field ($P > 0.05$). Both length and depth of lesions caused by *P. vulpinum* and *B. cinerea* were similar, *F. graminearum* caused slightly less severe symptoms, and *Geotrichum* sp. did not cause symptoms statistically different from the control. In trial 2, however, our results suggest that the interaction between CLS level, pathogen, and variety may affect sugarbeet rot depth ($P < 0.05$). There will be another timepoint at the end of the storage season, as well as a minimum of one mid-winter sample. Future trials will continue to examine the effect of CLS, as well as the varietal effect on rate of rot symptom development.

To measure the effect of CLS on respiration rate, roots of C-G333NT and HIL-9865 with high and low CLS ratings from trial 2 were stored in vented respirometry chambers at 42°F. Samples are taken periodically throughout the storage season to measure the beet respiration rate/kg. These beets are not inoculated with any storage pathogens. There was no significant difference in respiration rate of beets with high or low CLS levels following timepoint 1 and 2, 60- and 80- days post-harvest ($P > 0.05$). Additional rate measurements will be collected 100- and 120-days post-harvest. There does appear to be a difference in respiration rate among varieties. In addition to further evaluation of varietal effect on respiration, the impact of storage pathogens on respiration rate will be evaluated in the future.

References:

Fugate, K., Eide, J., Hakk, P., Lafta, A., and Khan, M. 2021. Impact of Cercospora leaf spot disease severity on sugarbeet root storage. Sugarbeet Research and Extension Reports (electronic publication).

Khan, M., Haque, M., Brueggeman, R., Zhong, S. Bhuiyan, M.Z.R., Poudel, R.S., Gross, T., Hakk, T., and Liu, Y. 2019. First Report of *Geotrichum candidum* Causing Postharvest Rot of Sugar Beet (*Beta vulgaris*) Roots in Minnesota and North Dakota. Plant Disease. 103(12):3278.

Campbell, L. G. 2015. PI 674103, *Beta vulgaris* L. subsp. *vulgaris*. U.S. National Plant Germplasm System. https://npgsweb.ars-grin.gov/gringlobal/accessiondetail?id=1923721.

McGrath, J.M. 2012. Germplasm releases: EL50/2; EL58 through EL66; SR99 through SR101 [CD-ROM]. 2012 Annual Beet Sugar Development Foundation Research Report. Denver, Colorado: Beet Sugar Development Foundation.

Smith, G.A. and Ruppel, E.G. 1971. Cercospora leaf spot as a predisposing factor in storage rot of sugar beet roots. Phytopathology. 61:1485-1487.

This work is supported by the Michigan Sugar Company, USDA-ARS, Beet Sugar Development Foundation, and Project GREEEN. We also thank Dennis Bischer, Corey Guza, and Michigan Sugar Company agronomists for their assistance in obtaining beet root samples.

Molasses Filtration with Automatic Discharge of Dry Solids

Stefan Schöpf*
Lenzing Technik GmbH
4860 Lenzing, AUSTRIA
(s.schoepf@lenzing.com)

ABSTRACT

Molasses contains approx. 60% of sugar. In order to extract this sugar and to increase the overall yield of a sugar mill, the molasses is processed in a chromatographic process into three phases. Whereas the natural betaine fraction is sold to third parties and the raffinate is added to sweeten animal feed, the sucrose extract is recirculated to the crystallization to recover additional sugar.

This three phase separation happens in a chromatographic process, which needs to be protected through proper filtration as the resin bed of the columns is rather sensitive and very costly to exchange.
Due to these facts precoat filtration using filter aid is the state of the art technology. As the sticky molasses is rather difficult to clean of the filter cloths usually a slurry discharge was used for regeneration, with the effect of efficient cleaning, but loss of product, respectively molasses.

In order to be able to process 100% of the molasses the Lenzing Technik GmbH has gone through an intensive trial phase together with the customer for more than a year to optimize the settings, parameters and consumables used at the Lenzing CakeFil equipment. These trials resulted in a dry solids discharge, molasses filtration process without product losses and additional recovery steps.

KEYWORDS

Cake filtration, precoat, molasses filtration, filter aid, filtration efficiency, molasses, dry discharge, efficiency improvement, increasing yield

Molasses Filtration with Automatic Discharge of Dry Solids

1. INTRODUCTION

Closing the loops and increasing yields are only two phrases which are often used, but need to be followed to be able to survive in a more and more competitive environment. These very general phrases also count for the sugar industry especially in regards of the changing market.

Thinking about closed loops and increased yields in sugar industry automatically Molasses comes into one's mind. Although it is almost everywhere already used in a sensible way, as e.g. addition to sweeten animal feed, it could be used to produce more sugar per beet and to gain higher margin from betaine fraction.
Therefore the molasses is processed in a chromatographic process into three phases. Whereas the natural betaine fraction is sold to third parties and the raffinate is added to sweeten animal feed, the sucrose extract is recirculated to the crystallization to recover additional sugar.

This three phase separation happens in a chromatographic process, which needs to be protected through proper filtration as the resin bed of the columns is rather sensitive and very costly to exchange.
Due to these facts precoat filtration using filter aid is the state of the art technology. As the sticky molasses is rather difficult to clean of the filter cloths usually a slurry discharge was used for regeneration, with the effect of efficient cleaning, but loss of product, respectively molasses.

In order to be able to process 100% of the molasses the Lenzing Technik GmbH has gone through an intensive trial phase together with the customer for more than a year to optimize the settings, parameters and consumables used at the Lenzing CakeFil equipment. These trials resulted in a dry solids discharge, molasses filtration process without product losses and additional recovery steps.

2. STATUS QUO

In the specific case this paper is targeting, a sugar mill is operating a Simulated-Moving-Bed (SMB) Chromatography for 70 000 tons of Molasses per year which the operator wants to increase to 100 000 tons with the parameters shown in Table 1.

The task of molasses filtration can be seen as an example only because chromatographic equipment always requires efficient filtration to provide a feed with a maximum of 5ppm suspended solids and a maximum particle size of 5µm. Within sugar refineries e.g. the recovery house can be replaced by chromatographic columns.[i]

Table 1: Operation Parameter for the described case

Molasses amount	70 000 T/a
Sugar Content	60 brix
Static viscosity	3800 cm/s
Dynamic viscosity	4.77 mPas
Operating Temperature	84 °C

During this capacity expansion project it turned out that filtration is the bottleneck and need to be increased in capacity.

2.1. Existing Filtration

Actually the molasses is filtered through two units plate type precoat filters with slurry discharge. One of these two units is always in operation the other one in regeneration. The slurry is recovered through a decanter unit to minimize product losses.

Figure 1: Left: existing plate type precoat filters / Right: Slurry decanter

As the performance of the plate type precoat filters is, additionally to the capacity problems, also unsatisfying in terms of filtrate quality an automatic Lenzing OptiFi® operated with 5µm woven screen is used for polishing afterwards.

Figure 2: Lenzing OptiFil® for downstream polishing

3. THEORETICAL BACKGROUND AND DEFINITIONS

3.1. Filtration Mechanism[ii]

In theory the retention of particles from fluids can be described in four idealized mechanisms, which are surface, depth, cake and precoat filtration. In the following lines these theoretical approaches are described. Although in reality always a mixture of these mechanisms appears.

- Surface Filtration

Surface or blockage filtration describes the mechanism, how solids block the pores of a filter media. This mechanism is most descriptive when soft and formable particles are filtered on a sieve. In this case the differential pressure over the sieve rises exponentially after the solids blocked the last free pore of the sieve, as shown in Figure 3.

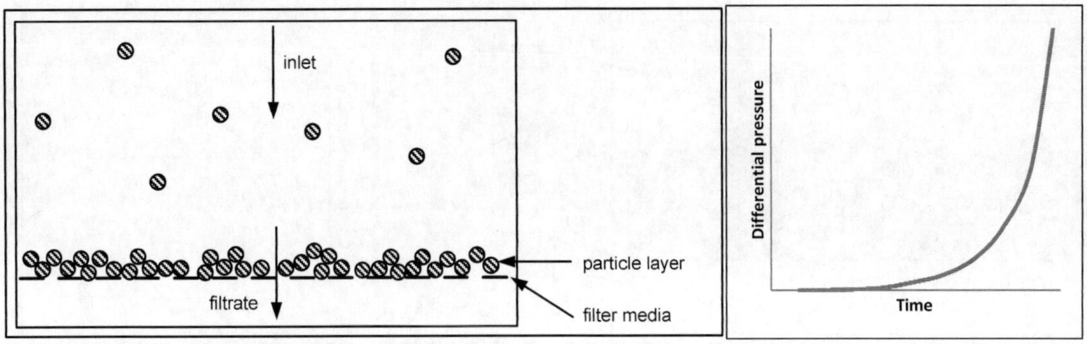

Figure 3: Principle of Sieve Filtration

- Depth Filtration

Figure 4 shows the principle of depth filtration, where solids are kept in the inside of a permeable multilayer filter media until the filter media has reached its absorption capacity. After that, the differential pressure increases either exponentially as already described above or the critical differential pressure leads to a break-through of the impurities into the filtrate.

Although the same principle applies for both, the single layers of the filter material as well as for the surface or blockage filtration, the dirt holding capacity is significantly higher.

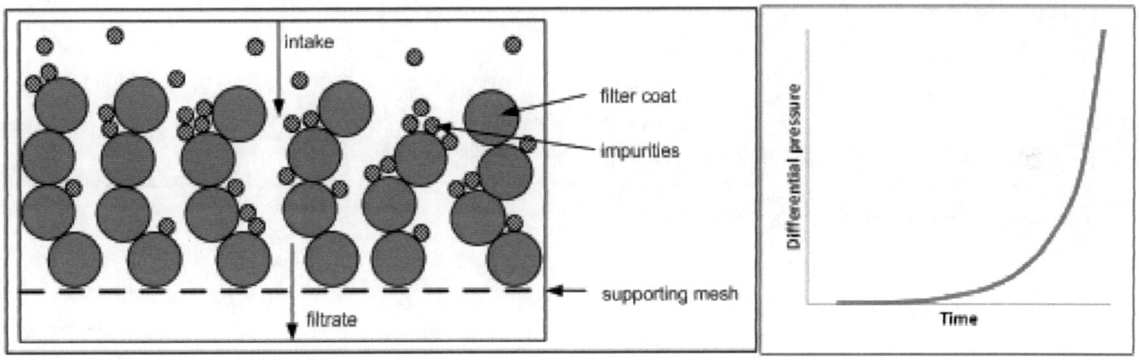

Figure 4: Principle of Depth Filtration

- Cake Filtration

During cake filtration the solid particles settle on the filter media and build up a filter cake of growing thickness. In the best case the resistance of the filter cake rises proportionally to its thickness. At a continuous flow rate, the differential pressure rises proportionally to the filtered volume.

As shown in Figure 5, typically the pores of the filter media have a bigger diameter than the solid particles. However the particles tend to form solid bridges above the pores. Until this happens it is possible that solid particles migrate into the filtrate.

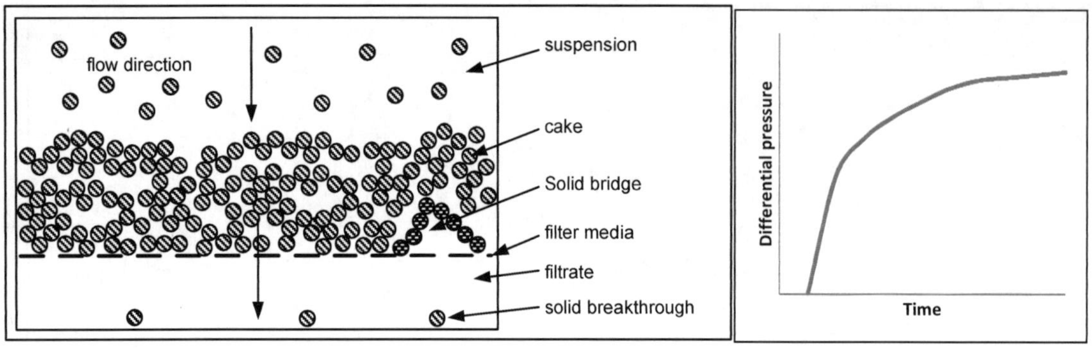

Figure 5: Principle of Cake Filtration

- Precoat Filtration

The theoretical background of precoat filtration is a kind of cake filtration. The only difference is that filter aid (typically Perlite, Diatomaceous earth or Cellulose) is added to the suspension to form a permeable filter cake.

This can be done as precoat and / or bodyfeed depending on the requirements of the filtration process and the characteristics of the particles.

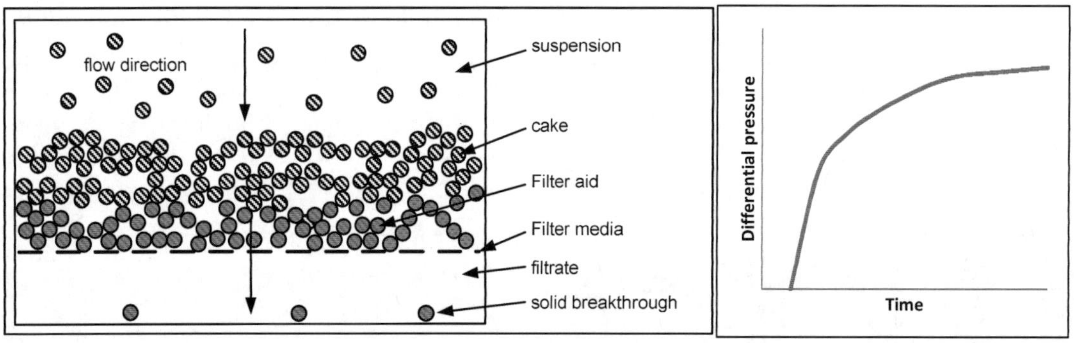

Figure 6: Principle of Precoat Filtration

3.2. Filter aids

As the impurities contained within the molasses, do not form a permeable filter cake it is necessary to add filter aids for proper filter cake formation. For Molasses Filtration the filter aid of choice is Perlite, but for varying filtration tasks varying filter aids are required. In the following lines the most prevalent ones are described.

- Perlite

Perlite is a naturally occurring volcanic glass which thermally expands upon processing. Perlite is chemically a sodium potassium aluminum silicate. After milling, a porous, complicated structure is present, but because its structure is not as intricate (or tortuous) as that of diatomite, perlite is better suited to the separation of coarse microparticulates from liquids having high solids loading. Perlite is lower in wet cake density than diatomite and this enables using less filter aid (by weight).[iii]

- Cellulosic

Cellulose filter media is produced by the sulphite or sulphate processing of hard woods. Cellulose is characterized by its high aspect ratio, which enables it to precoat a septum very easily. It is most often used in that capacity in combination with diatomite. Like perlite, cellulose possesses a less intricate structure than diatomite. Attempts have been made to add structure. These include fibrillating the strands. Cellulose also has the ability to operate in elevated pH environments above 10, making it frequently used in the chlorine-caustic industry to filter the brine feed to electrolysis membrane separators. Another application of cellulose is in treating machining oils and cutting fluids, to break the emulsion or to trap metal fines. Cellulose can be burned out after the filtration to recover the metal particles.

- Diatomite

Diatomite is obtained from diatomaceous earth, a sediment greatly enriched in biogenic silica in the form of the siliceous frustules of diatoms, a diverse array of microscopic, single-cell algae of the class Bacillariophyceae. These frustules are sufficiently durable to retain much of their structure through long periods of geologic time and through thermal processing. Diatomite products are characterized by an inherently intricate and highly porous structure composed primarily of silica, along with impurities of alumina, iron oxide, and alkaline earth oxides.

- Activated Carbon

Activated carbon's range of applications extends from removal of various impurities, e.g. flavours and odours, to removal of disinfectants and their decomposition products from water. In addition, it maintains long-term colour stability. The structure and concentration of the impurities to be removed determines subsequent treatment. To ensure efficient absorption, the pore size should be almost equal to the size of the molecules which are to be processed. A broad range of grinding degrees is available and the user-specific activated carbon grades determine the purity of the liquid's components, such as water, carbon dioxide, and sugar. This approach ensures all quality criteria can be met.*iv*

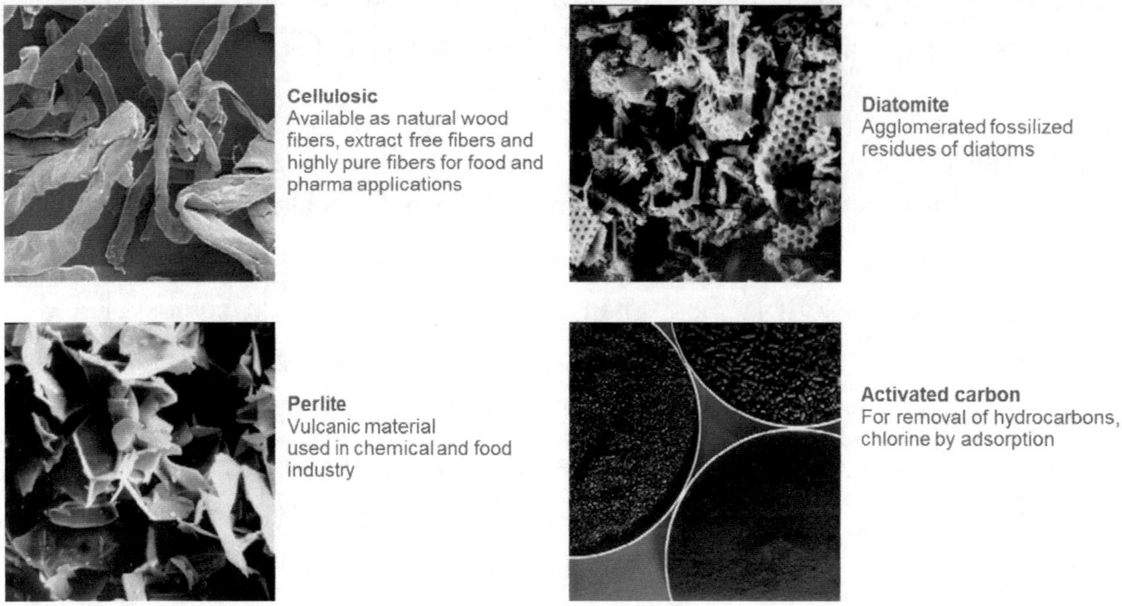

Figure 7: Overview on Filter aids available

Figure 7 shows microscopic photographs of different filter aids where e.g. it is visible that Perlite has a higher specific surface then Cellulosic, but less then Diatomite filter aids.

3.3. Filter Material/ Cloth

Besides a wide range of filter aids available also numerous options of filter materials can be applied. Figure 8 gives an overview on those. Varying applications require varying attributes of the filter cloths. Choosing the best one for each task required experience and knowledge.

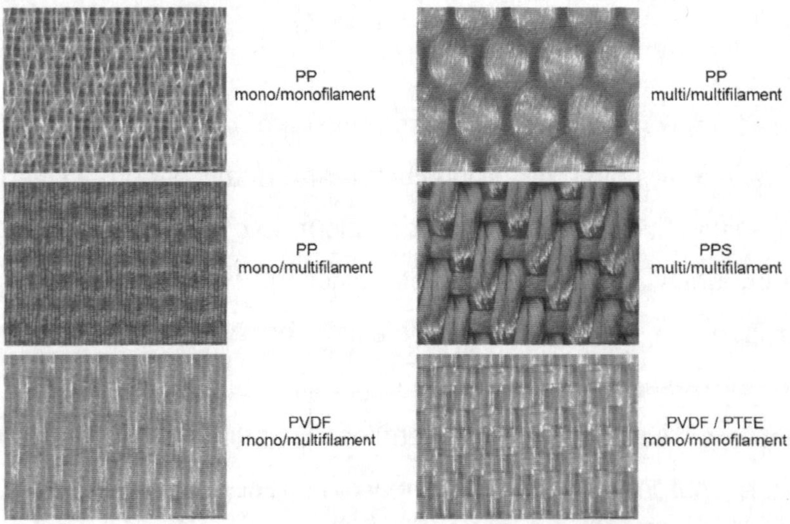

Figure 8: Range of filter material available Lenzing CakeFil

4. LENZING CakeFil

4.1. Function and Design

Figure 9: Left: outside of filter vessel CakeFil / Right: inside of filter vessel CakeFil

During the filtration, which takes place in a pressure vessel, as shown in example above in Figure 9, the liquid is pressed from the outside through the filter medium. Solids collect on its surface and form a uniform cake, which remains on the filter elements due to a continually maintained pressure differential across the filter candles.

Besides earlier mentioned parameters like filter aid, filter material and the vessel design, one of the keys for an economic candle filter is the choice of the correct candle design. According to the characteristics of the filter cake the type of discharge (slurry or dry) the candle design has to be adjusted accordingly. Figure 10 shows an excerpt of the candle range of Lenzing Technik GmbH as an example.

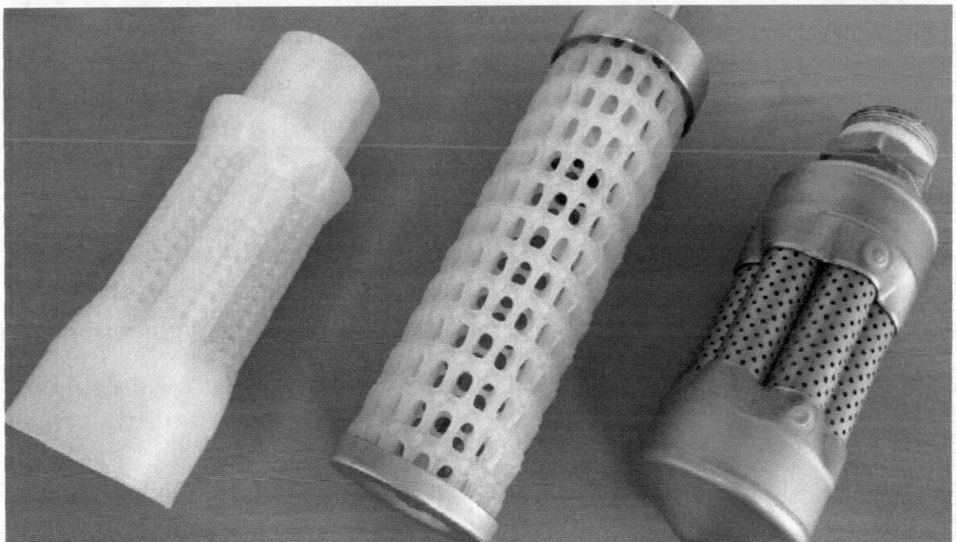

Figure 10: Excerpt of available candle variations

4.2. Step Sequence

With this kind of equipment the filtration is divided into multiple phases.

Starting from a clean filter material, in our case a filter cloth, as shown in Figure 11 one filtration cycle can be divided in different sequences.

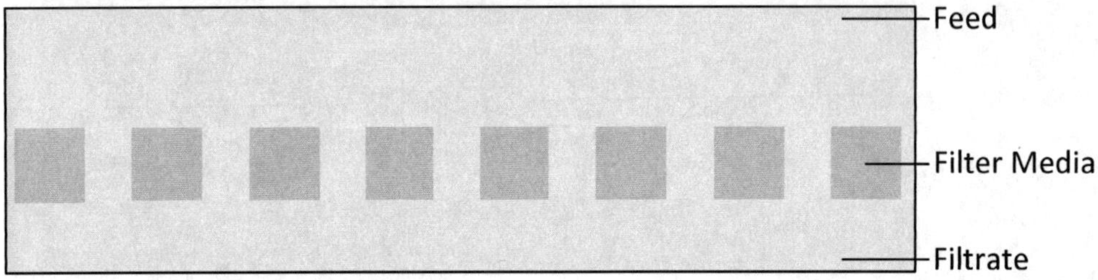

Figure 11: Clean filter material

- Precoat

During precoating the filter vessel is filled with a mixture of filtered fluid or water and a comparably coarse filter aid which is in this case a Perlite. The precoat layer is necessary to provide the support and depth for the fine filter cake created during filtration sequence. Figure 12 shows the bridge formation through a relatively coarse filter aid above the pores of the filter cloth.

Figure 12: Precoat phase

During the formation of the precoat layer the fluid is recirculated to the feed line to prevent the migration of particles into the filtrate line.

- Filtration

Whilst the first step was only the preparation of the filter media, at this stage the main filtration starts. A mixture of suspension containing impurities and sometimes a finer filter aid dosed as a bodyfeed is fed onto the precoated filter material. Whilst the clean liquid is passing the filter cake, the impurities and the filter aid particles are kept on or in the depth of the precoat layer.

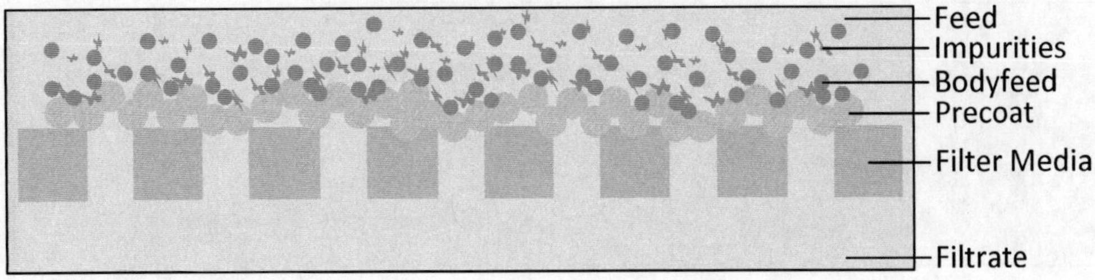

Figure 13: Filtration phase

- Cake washing

As soon as the end of a filtration sequence is reached the filter vessel gets emptied (filling volume is recirculated to the feed tank). To maximize the sugar yield the cake afterwards gets washed with water through spray nozzles. This washing water is afterwards also recirculated to the feed tank and used for dilution of the molasses.

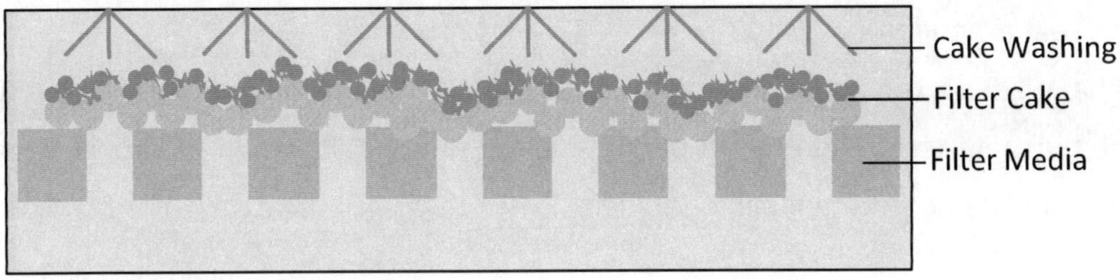

Figure 14: Cake Washing phase

- Cake drying

The washed cake is then dried with pressurized air to prepare it for the blow off. This step makes the filter cake more brittle.

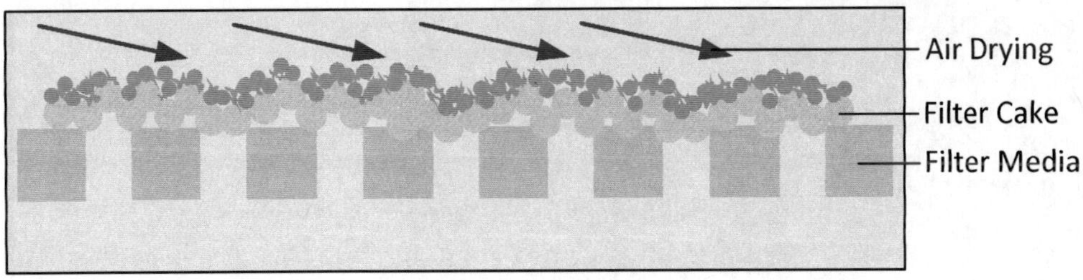

Figure 15: Cake drying

- Cake discharge

Afterwards the filter cake is blown off in reverse filtration direction by a strong pressurized air impulse which breaks the brittle filter cake and allows for the gravity driven discharge of the dry filter cake through the bottom.

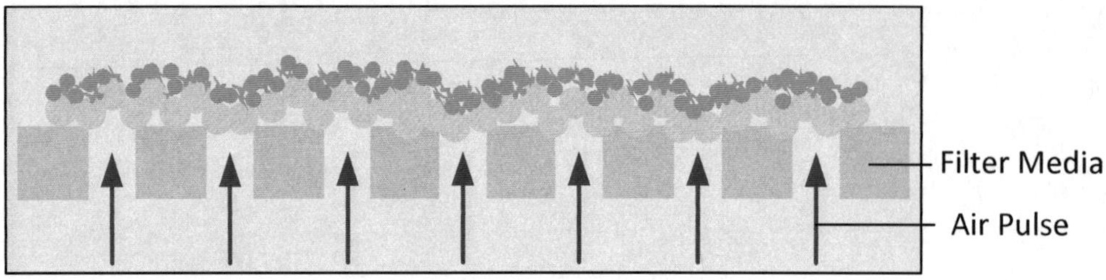

Figure 16: Cake discharge

Optionally after the discharge there can be a sequence of cloth washing before the sequence starts from the beginning.

5. Test execution and Results

5.1. Test execution

The test have been executed in a bypass stream of the molasses feed of the existing plate type precoat filters. Therefore a plug and play pilot unit shown in Figure 17 was integrated with hose connections into that line.

Figure 17: Pilot unit Lenzing CakeFil in operation

The setup allowed an independent operation with varying filter aid concentrations and the full range of step sequences to try the best configuration. Table 2 shows the configuration of the three most promising trials.

Table 2: Configuration of most promising trials

Trial	Filter material	Precoat [kg/m²]	Bodyfeed [g/L]	Cake washing	Cloth washing
01	LT-FM-02-13-03-007	0,83	1	No	No
02				No	Yes
03				Yes	Yes

5.2. Results

To sum up the executed trials only the three above mentioned trials will be thematised in the following lines as the full range of trials would be too extensive.

- Trial 01

This trial has shown that the quality parameter can easily achieved with these concentrations and filter material set up. Furthermore it has shown that a cyclical cloth washing is required to enhance the performance of the filter material.

- Trial 02

The trial has shown that the sequence of cloth washing had the desired effect in terms of an enhanced cloth performance and life time. As a next parameter the rest sugar content within the filter cake was observed and it showed that without cake washing the rest sugar content was 41.9 which was not acceptable for the customer and led to an additional test round.

- Trial 03

With the cake washing the rest sugar content of the dried filter cake could be minimized to a level of 2.4 which is not only acceptable, but an improvement to the actual situation.

6. Conclusion

The outcome of the tests showed that 2 units of Lenzing CakeFil each having 29m² of filter area can replace the actual setup of two times plate type precoat filters plus automatic filter for polishing as well as the decanter for the slurry recovery.

Figure 18 shows the existing filtration compared to the Lenzing CakeFil concept in a simplified form.

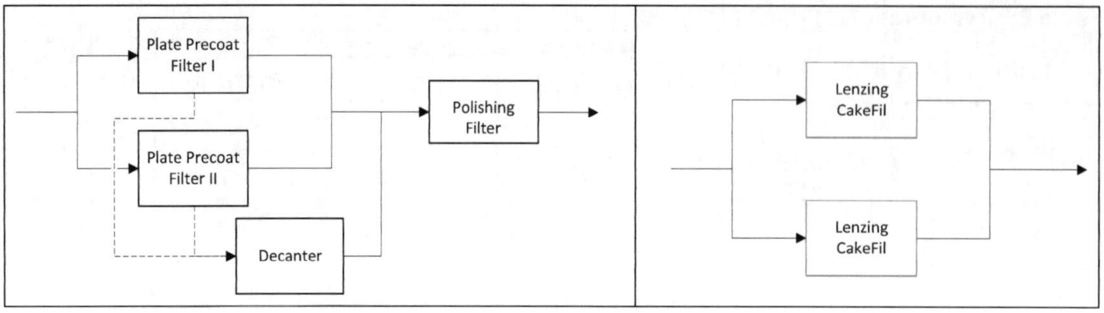

Figure 18: Left: Existing Filtration / Right: Lenzing CakeFil concept

In fact the tests have proven that an existing three stage filtration could be replaced with a single stage. Whilst achieving same levels in terms of quality and even slightly less filter aid consumption and therefore operating cost.

In Table 3 the average time per sequence is shown. One total step sequence takes 155 to 229 minutes whereas the filtration lasts for 120 to 180 minutes. This is a typical plant design with roughly 75% effective filtration and redundant filter vessels to provide reserves in terms of quality as well as capacity.

Table 3: Average time for one step sequence

Sequence	**Average Time**
Filling	5'
Precoating	15-20'
Filtration	120-180'
Emptying	5'
Cake Washing	3-5'
Cake Drying	2-5'
Cake Discharge	2-5'
Cloth Cleaning (cyclic)	3-4'

---

i M. Kearney, T. Pryor, L. Velasquez, A. Hieb, B.-C. Schulze; 71st SIT: **A Novel Approach to Raw Cane Sugar Refining using Ion Exchange and Chromatography**

ii S.Schöpf; SIT2015; **Efficiency improvement through unique automatic backwash filter**

iii Thomas E. Sulpizio; 1999; **ADVANCES IN FILTER AID AND PRECOAT FILTRATION TECHNOLOGY**

iv https://www.vulcascot.at/activatedcarbon

Molecular profiling of the sugar beet pathogen *Fusarium secorum*

Subidhya Shrestha [a,b], Roshan Sharma Poudel [a], Ronnie de Jonge [c], Kimberly Webb [d], Gary Secor [a], Melvin D. Bolton[a,b,†]

[a] Department of Plant Pathology, North Dakota State University, Fargo, ND 58102

[b] U.S. Dept. Agriculture – Agricultural Research Service, Northern Crop Science Laboratory, Fargo, ND 58102

[c] Plant-Microbe Interactions, Department of Biology, Science4Life, Utrecht University, Utrecht, NL

[d] U. S. Dept. of Agriculture – Agricultural Research Service, Soil Management and Sugar Beet Research Unit, Fort Collins, CO 80523

[†] Corresponding author: M. D. Bolton; melvin.bolton@usda.gov

Fusarium yellowing decline caused by the fungal pathogen *Fusarium secorum* has become an emerging problem in the sugar beet industry. Since *F. secorum* is not closely related to other *Fusarium* pathogens of sugar beet, little is known of the virulence mechanisms of this pathogen. To that end, we utilized whole-genome sequencing of the pathogen, xylem sap mass spectrometry, and transcriptome analysis of *F. secorum* infected-sugar beet plants to understand the molecular basis of virulence of this pathogen. *Fusarium secorum* showed an increased genome size due to an increased number of introns and repetitive elements. Additionally, we successfully developed a CRISPR-Cas9 ribonucleoprotein mediated gene-editing technique to disrupt target genes encoding candidate effector proteins in the pathogen. There were five gene targets, and among them, one gene target (*Fsec2*) was identified as a virulence factor of *F. secorum*. This study provides valuable genomic resources and a better understanding of the virulence strategies of an important pathogen of sugar beet.

THOMAS, LACIE*, JAIME WILLBUR, DANIEL BUBLITZ, and KURT STEINKE. Department of Plant, Soil, and Microbial Sciences, Michigan State University, Plant and Soil Sciences Building, 1066 Bogue Street, East Lansing, MI 48824. **Utilizing Foliar Boron for Managing *Cercospora beticola*.**

One of the more severe foliar pathogens capable of causing damage to sugarbeet is *Cercospora beticola*, the causal pathogen of Cercospora leaf spot (CLS). When not managed appropriately, CLS can reduce sugarbeet yield 40%. Although plant defoliation caused by the disease directly impacts root size and sugar quality, other factors including leaf regrowth and impurities within the root affect plant health and crop quality. Managing CLS remains difficult due to resistance to several fungicides. Management strategies including boron-containing compounds have been discussed as possibly containing fungistatic properties with the ability to reduce disease severity in the field. Field studies were established to investigate the effects of foliar applied boron on sugarbeet plant health and CLS disease severity. Treatments included a standard fungicide program, three foliar boron treatments (0.1, 0.25, or 0.5 lbs. B per acre) applied at 10-day intervals for 7 total applications without a standard fungicide program, three foliar boron treatments (0.1, 0.25, or 0.5 lbs. B per acre) applied at 10-day intervals for 7 total applications in conjunction with a standard fungicide program, and a nontreated check for a total of 8 treatments. Measureables collected include plant tissue nutrient analysis, percent canopy coverage, canopy normalized difference vegetation index (NDVI), visual disease incidence and severity, sugar quality, and yield. In addition, complementary *in vitro* studies will be conducted to test boron efficacy on pathogen growth. Preliminary results from the 2020 growing season will be presented and discussed as part of an integrated CLS management program.

AUTHOR INDEX